Marvellous Mammoths

Mammoths in Huntingdonshire

Huntingdonshire Local History Society
Goodliff Award

Dr Chris Thomas

Norris Museum Volunteer

Milton Contact Ltd
3 Hall End, Milton, Cambridge, UK
CB24 6AQ

www.miltoncontact.co.uk

How it all started

A small, insignificant, ginger tangle of fur next to a giant tooth and tusk. "Mammoth Hair" and "...40,000 years ago" leapt out from the label.

Now, I had looked at hair from dogs, cats, mice, rabbit, wolf, lion, cheetah, giant anteater, brown bear and polar bear and even a walrus whisker, amongst others.

40,000 year old mammoth hair!

I'd only wandered into the Norris Museum in St Ives to kill some time between business meetings.

40,000 year old mammoth hair!

I turned to Helen Giles, the museum's curator in 2013. Feigning nonchalance, I asked "Have you ever looked at the mammoth hair under the microscope?"

"No."

"Would you like to?"

"YES!"

And so, I was drawn into the friendly family of the Norris Museum and its volunteers.

.....

Early in 2014, Helen turned to me and said, "I think we should have a mammoth exhibition. Would you like to help organise it?"

A year on, with Assistant Curator Gilly Vose's help, a story emerged.

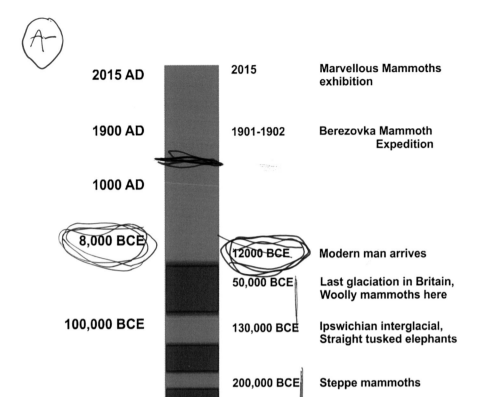

2015 AD	2015	Marvellous Mammoths exhibition
1900 AD	1901-1902	Berezovka Mammoth Expedition
1000 AD		
8,000 BCE	12000 BCE	Modern man arrives
	50,000 BCE	Last glaciation in Britain, Woolly mammoths here
100,000 BCE	130,000 BCE	Ipswichian interglacial, Straight tusked elephants
	200,000 BCE	Steppe mammoths
	300,000 BCE	Early human tools found Steppe mammoths
	800,000 BCE	Early human first recorded in UK
1 million BCE	2,500,000 BCE	Southern Mammoth
	2,580,000 BCE	Current Ice Age begins
10 million BCE		
	66,000,000 BCE	Cenozoic Era begins 'The Age of Mammals'

Figure 1. Timeline of events in our region of the UK

Contents

Overview

Join me in this booklet, as I retrace the steps that began with my first contact with the Norris Museum's mammoth hair.

After establishing that the hair looked genuine, I asked myself, "How did it get to the Norris Museum?" An insignificant-looking envelope offered me a link to the Berezovka Mammoth in the Museum of Natural History in St Petersburg. It was the first – and most famous – intact specimen to be found.

The discovery and recovery of the Berezovka mammoth is an unexpected epic and adventurous link to our mammoth hair sample. In winter 1901, scientists set out on an epic 18,000 km round trip to Siberia. Travelling by train, boat, horseback and dog-sleigh, they successfully returned with the mammoth after nine gruelling months. It is still a centrepiece in the museum in St Petersburg.

So, where did mammoths come from – and could they have lived in and around Huntingdonshire in the UK?

We find out that for the past two and a half million years, our planet has been in an ice age, that within an ice age we have a 100,000-year rhythm with periods of intense cold, where ice sheets invade our country. These are interspersed with warm periods and, in fact, we are currently in the most recent warm interglacial.

And yes, we not only had one type of mammoth but three, with an elephant species thrown in, living here in St Ives, Huntingdonshire and the UK.

Humans also came across the land-bridges that regularly linked us to the European mainland when sea levels fell. *Homo heidelbergensis* lived on these shores 800,000 years ago and would certainly have seen the Steppe mammoth or even the Southern mammoth – as well as other unusual animals. Modern humans *Homo sapiens* returned to these isles about 40,000 years ago after a gap in human occupation. We have lived and prospered here ever since.

We have the evidence as fossils and tools in the collection of the Norris Museum.

Every artefact in the Norris Museum has a story. A seemingly insignificant bundle of dull brown hair has a particularly far reaching one, both in links to the other side of the world and deep into the past.

I had an enjoyable year discovering these stories and gaining new knowledge, culminating in an exhibition, "Marvellous Mammoths" at the Norris, and this booklet.

I hope you will be amazed and enjoy the journey too.

Is it Woolly Mammoth Hair?

We'll take a look at the investigation of the mammoth hair sample at the Norris Museum and the results that confirmed it to be genuine. I studied the hair by eye, then under different microscopes and compared it with other animal hairs I had looked at in the past.

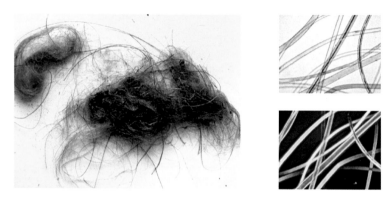

Figure 2.The Norris Museum mammoth hair. Left: The clump of hair as displayed. Right: Hair under the microscope with transmitted light (top) and using polarisation filters (bottom). The Norris Museum

I arrived at the Norris Museum around 10am, on the 8th October 2013. I still remember the thrill of anticipation as the display cabinet was opened and the card with the hair was carefully removed and transferred to the small room behind reception where the volunteers work.

As the sample had to stay in the museum, I'd done some careful planning. The strategy was to prepare any samples if needed, do a visual study with basic notes, and take lots of photographs. I could then work on the photographs at leisure at home.

So what do you need to look for when studying hair? What did we discover?

First, here are some important details of animal hair that need to be considered and are specific to each animal:

- Does the sample include full lengths of hair or fragments?
- Hair width and cross-section
- Surface scales
- Colour
- Internal structure
- Condition (intact or decaying).

Full length or intact?

A trichologist (an expert on hair) will tell you that it is important to have a complete hair, from the bulb at the base of the hair, through the whole of the shaft to the hair tip.

Unfortunately, with our sample, I could not see any hair bulbs. The ends of the hair I did see were either broken or frayed. Did this mean that the hair had been cut off? Or, had structures decayed over time? We do not know.

Hair width and cross-section

Looking at the mammoth hair on its card under a low power microscope, I could see that the fibres varied from very fine to thick. They all seemed to be round in cross-section. Natural hair does vary in width. In contrast, man-made fibres are much more uniform.

On average, the hairs were about 0.05mm wide, which matched measurements by other researchers. However, if you look at the chart of hair widths, you can see that there are thicker hairs, some reaching to almost 0.4mm in diameter (Figure 3).

This is unusual. Most animal hairs have quite a narrow range of widths. You can see this with the human and horse-tail hairs on the same chart. Where animals have a fine undercoat and thick overcoat, as in your pet cat or dog, the fine and thick hairs also split into distinct size groups.

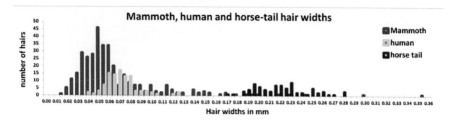

Figure 3. Mammoth hair widths (red) compared to human hair (yellow) and horse tail hair (blue)

A possible explanation for the variation in width lies in the origins of mammoths (see the chapter on Mammoths and man in St Ives). Their early ancestors originally came from tropical Africa. Like modern day elephants, they did not need hair to keep them warm. They probably lost the thin insulating undercoat hair and only had a sparse covering of coarser hair.

M. L. Ryder (Nature, 1974) looked at mammoth skin and the pattern and sizes of the hair follicles. He suggested that when woolly mammoths adapted to the cold of the ice age, they re-developed an insulating undercoat of fine hair from the coarse ones. The wide range of hair widths I found could reflect this adaptation.

Surface scales

Animal hair has an outer layer of scales, called the cuticle. When lit from the side, you can just about make out the scales on most hair.

On our mammoth hair, I could only see the occasional hints of scales (Figure 4). Most of the mammoth hairs were smooth. Some mammoth hair found by other research groups also had surface scales. Perhaps the cuticle had been degraded over the passage of time.

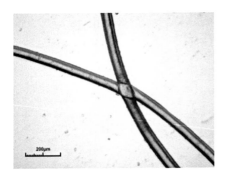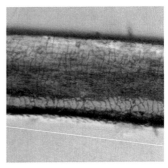

Figure 4. The surfaces of fine mammoth hairs on the left and a human hair showing surface scales on the right. The Norris Museum

With the naked eye, the mammoth hair looks dull brown. The closer you get to look at the hair, the more transparent most of the hairs appear.

Coloured human and animal hairs look different to our mammoth hair under the microscope. With other animal hair, colour is concentrated in little spots within the hair shaft. The darker the hair, the more pigment spots. In the mammoth hair, the colour is not in spots but seems to be an even tint throughout.

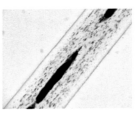

Figure 5. Left: Mammoth hair; Centre: Light human hair; Right: Dark dog hair. Note that the mammoth hair lacks the pigment spots seen in the human and dog hair

This is a common feature of mammoth hair. The browns of any mammoth hair found are thought to be due to environmental staining or degradation of the hair itself. We do not know what colour mammoths were, though Mammoth DNA studies show that both light and dark coat colours were possible. Figure 8 shows a particularly good example of surface staining of a mammoth hair.

Internal structure

Apart from the outer cuticle, the inside of a hair is called the cortex. In many animal hairs, there is also an air filled region called the medulla. The medulla can be quite distinctive for certain species – as you can see from the patterns of the medullae in the different animal hairs in figure 6.

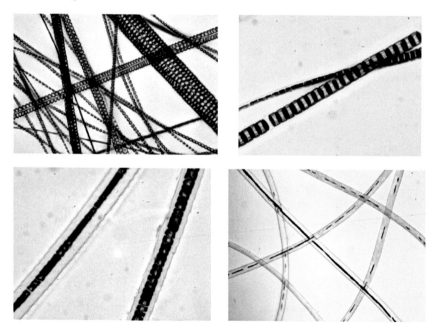

Figure 6. Different animal hairs under the microscope. Top row: Rabbit and field vole. Bottom row: Wolf and alpaca

The air-filled medulla can act as additional insulation. In fact, hollow fibre duvets and arctic clothing copy this effect.

Most of our mammoth hair does not have an air-filled medulla. Only the occasional hair shows one, as in figure 7.

Again, this is a known characteristic of mammoth hair.

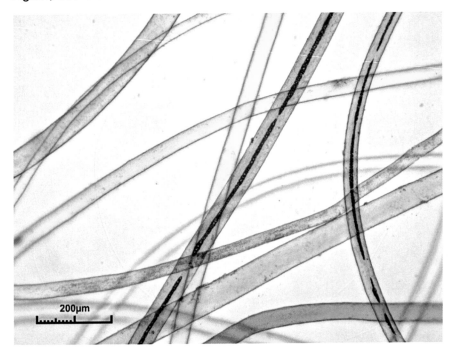

Figure 7. Mammoth hair. In contrast to the other animal hairs in figure 5, only the occasional hair has a medulla. The Norris Museum

Hair condition

The mammoth hairs looked quite intact. They were still quite strong when handled, though at higher magnification some thicker hairs appeared glassy and brittle.

This is actually quite amazing. Hair is made up of a protein called keratin and protein is a food source. Hair might be too tough for us to eat, but bacteria and fungi will attack hair and digest it, given a

8

chance (see figure 8). Under normal conditions, the mammoth hair should have decayed long ago.

Two things can prevent decay; extreme dryness or freezing. Since this is woolly mammoth hair, preservation by freezing is most likely.

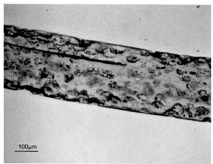

Figure 8. Left: Norris museum coarse mammoth hair, showing smooth surface. Riaht: Another Siberian mammoth hair showing signs of surface degradation

Conclusion

The mammoth hair sample in the Norris Museum collection and display is consistent with observations made by other scientists on other woolly mammoth hair samples. It is the right average width; only a few hairs have a medulla; the hair colour seems to be due to staining from the environment or degradation. The otherwise good condition suggests that it was preserved, possibly in a very dry area or frozen in ice.

We do find lots of mammoth bones and teeth in our region. But the last permanent ice and glaciers left Britain thousands of years ago. So where did our mammoth hair come from?

The answer lies with the mysterious envelope used to donate the mammoth hair to the Norris Museum.

Investigator Guide to Studying Hair

Hair is a great subject for anyone to study; there is so much of it around. You can find it on people's heads; coming off from your pets when you stroke them; in your clothes as wool from sheep or alpaca; or even when you visit the countryside - on fences or posts where wild animals have pushed through.

A comprehensive guide to investigating hair can be downloaded here:

https://dl.dropboxusercontent.com/u/1646983/Mammoth%20booklet/investigator-guide-to-studying-hair.pdf

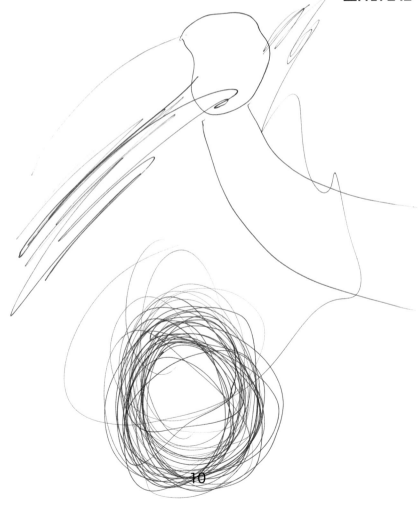

The Mysterious Envelope

The mammoth hair exhibit at the Norris Museum had arrived in an envelope which was initially lost amongst the museum records. The rediscovery of the envelope and the detective work to reveal the wealth of information hidden within it are described. The make of the envelope, the style of writing and pen used and of course the messages all pointed to the sample being taken and recorded in the envelope between 1902 and 1914.

The Norris Museum mammoth hair sample was simply recorded as:

Record X.1559, Paleontology, Russia, Berezovka. Strands of Mammoth hair. A tuft of ginger hair, with pieces of short under-hair, and longer pieces of coarse outer hair.

It had been with the museum for decades. Helen Giles, the curator in 2013/14 remembered a story that the hair had come in an old envelope that explained the hair's origin. Since this was before her time and there was no record of the envelope, it could have been anywhere in the museum's large store.

Volunteers at the Museum are working through the museum's stored collection and old records to update the catalogue. Many finds are just coming back to the light of day after lying in their boxes for nearly a century. It could have been decades before the envelope finally resurfaced.

We then had a stroke of luck. There was a party at the museum and the previous curator, Bob Burn Murdoch, had come along. He had served the museum for 32 years until retiring in 2012 and therefore knew the Norris and its contents well. We asked Bob if he could remember the mammoth hair and where the envelope might be.

"It's probably in a box labelled 'Displayed Objects'. That's where I kept odds and ends relating to objects put on display. It used to be on that shelf over there."

Bob pointed to an empty shelf. I looked at Helen, "We tidied those into the library!"

I left the party and dashed into the library, looking at every stack of museum boxes, and there were many.

Within minutes, I found one labelled 'Displayed Objects'. Opening it, there was the envelope we had been looking for.

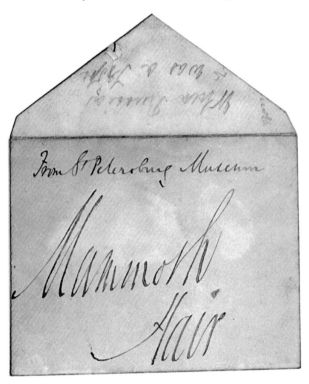

Figure 9. The front of the mysterious envelope, claiming to contain mammoth hair from St Petersburg Museum. The Norris Museum

Detective work on envelope and sample

The envelope held clues both to the likely date the mammoth hair had been collected and its origin. The clues lay in the envelope itself, the style of writing upon it, and, of course, what was actually written.

The Envelope

The envelope, when sealed, was rectangular and about 4¾" by 3¾" (about 120 mm x 96 mm) in size.

The symmetrical and clean cut edges of the traditional envelope show that it was machine cut and folded. People used to cut out their own envelopes until the mass production of envelopes began in 1845, with the patent by Edwin Hill and Warren De La Rue. For the next 50 years, these were cut out diamonds or lozenges that were folded to create an envelope. It was up to the letter writer and sender to then glue it as they saw fit.

There is evidence of glue on the opened flap, but none seems to have spread beyond the other glued down edges. This might suggest a pre-gummed/pre-glued envelope. These first came into production shortly before the 1900s.

Writing style

The writing on the envelope could be divided into 3 elements:
1. The letters in ink on the front
2. Writing in ink on the inside of the envelope flap
3. Writing in pencil all around the back of the envelope.

The ink lettering looks as if from an older age, with a copperplate style that dominated English handwriting from the 19th through to the mid-20th century.

Interestingly, if you home in on the letters, you can see that they were written with an old flexible nib. The wider down-strokes of the letters are created by pressing down slightly harder on the pen. The two tines of the flexible nib separate, giving a wider ink line. You can actually see the individual tine lines on either side of the wider stroke if you zoom in. This stroke then smoothly goes over to a thinner line as the pressure is reduced for the upstroke.

Flexible nibs were replaced with firmer nibs in fountain pens by the 1940s. This either gives a uniform line or, if you have an italic or broad nib, a line that changes between thick down-strokes and thin upstrokes in a different way to the flexible nib.

The writing in pencil on the back is in a similar handwriting, but currently we cannot say more about it.

We have narrowed the time the envelope was written to between 1900 to 1940.

The texts

Front of the envelope
"From St Petersburg Museum." St Petersburg was the capital of Russia between 1713 and 1918. Its name changed to Petrograd on 1st September 1914 (after the outbreak of the First World War). It only recently regained its old name in 1991.

"Mammoth Hair." Identifies the sample.

Inside flap of envelope
"Found in the ice Siberia."

Back of envelope
"I M C Nottage took this hair off the mammoth in the museum St Petersburg." The name could have been Chottage, Hottage, Nottage, Noltage or even Hollage.

My search of the 1911 census found one James Hottage, one Eliza Hollage and 799 Nottages. (Nottage comes from old English/French 'Nuthatch').

- If the sentence is "I, M. C. Nottage", there were 58 men and women called M Nottage in 1911
- If the statement was by "I. M. C. Nottage" , there were 6 men and women in 1911
- If it was by J. M. C. Nottage, there were 41 men and women listed as J Nottage in 1911.

I could not really identify a Nottage who might have gone to Russia to see the mammoth in the museum. But then – see *Hot Press* below.

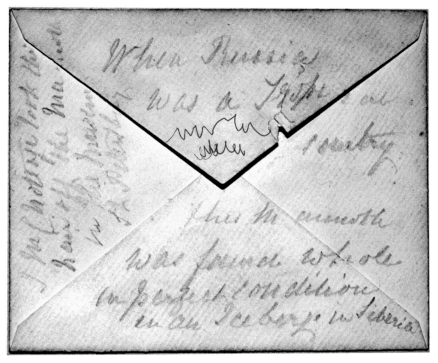

Figure 10. The reverse of the envelope with numerous texts in pencil. The Norris Museum

"When Russia was a tropical country." There was an old belief that mammoths lived in the tropics and were killed by a catastrophic freeze. Indeed, you can still find websites on the internet that follow or try to explain this theory.

"This mammoth was found whole in perfect condition in an iceberg in Siberia." Until the 1960's, the only intact mammoth found was the Berezovka Mammoth. Found in 1901-1902, it has been exhibited in the Zoological Museum in modern St Petersburg up to the present day.

Conclusions

The following facts suggest that this sample of mammoth hair was collected between 1902 and 1914 by someone with the surname Nottage, during a visit to the St Petersburg Museum:

- The envelope
- The handwriting and penmanship
- Use of the name "St Petersburg" which changed to Petrograd in 1914 with the outbreak of WWI
- The fact that the *first intact mammoth ever found* was not exhibited until 1902, in the Natural History Museum in St Petersburg.

Whether it was as a result of a scientific collaboration or an overambitious souvenir hunter/museum vandal, we do have a connection to the first and most famous intact woolly mammoth find; The Berezovka Mammoth of St Petersburg.

Hot Press! Nottage found!

Huntingdonshire - St Petersburg connection

Early in March, Linda Reed visited our Mammoth exhibition in the Norris and revealed that there were active links between St Neots and St Petersburg in the past. Craftsmen from our region regularly travelled to St Petersburg to work there during the 1860 to 1890. Whilst Linda could not find any Nottages in her records, she also came across across the mention of a William Hugh Nottage Mumford, who in the 1911 census was a undergrad of Cambridge University, aged 21 living at The Croft, Houghton. He later became Assistant Minister at St Ives Church between 1913-17.

Martha Christiana, Lady Nottage

Jenni Bishop looks after the Norris Museum's Twitter account and picked up a vital clue by Wendy Stacey. Wendy tweeted that she enjoyed the exhibition and thought that our missing Nottage might be Martha Christiana Nottage, wife of Lord Mayor George Swan Nottage, who died in office in 1885.

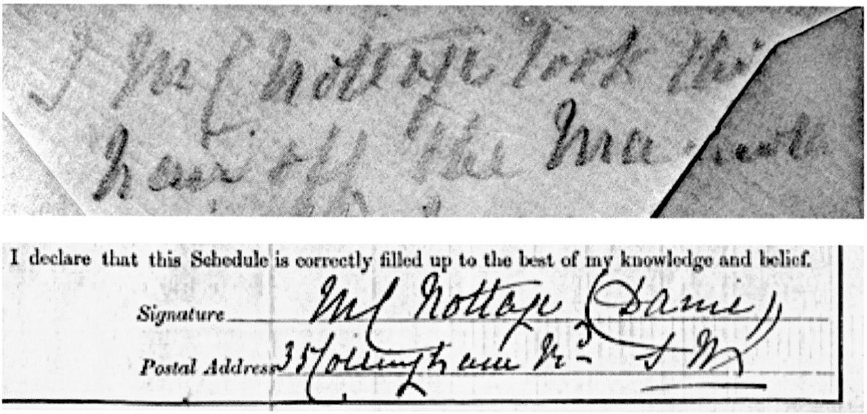

Figure 11. Comparison of envelope signature and census signature of Lady M C Nottage

Lesley Akeroyd, Museum Assistant and Archivist, picked up the conversation with Wendy at Cambridge University Library. Wendy had found a copy of the 1911 census form, signed by Lady Nottage. The signature appeared identical to that in the phrase on our envelope "I, M. C. Nottage".

The research continues to see if we can establish the link to St Petersburg through other records of Lady Nottage's life.

Investigator Guide – Writing

Writing, writing materials and surfaces proved a wealth of opportunities for your own research.

A four page guide to investigating writing can be downloaded here:

https://dl.dropboxusercontent.com/u/1646983/Mammoth%20booklet/investigator-guide-writing.pdf

Rescuing the Berezovka Mammoth

How the first complete mammoth was discovered on the banks of the Berezovka River in Siberia and the 'mammoth' journey undertaken in 1901/1902 to recover it for scientific excavation and public display by the Russian Academy of Sciences in St Petersburg.

Spring 1901 and the Russian Academy of Sciences in St Petersburg was agog with the news that an intact mammoth had been discovered on the banks of the River Berezovka in Siberia.

The last one had been found by a chieftain of the Tungus people in Eastern Siberia, in 1799. However, he kept it a secret, waiting till the tusks came loose.

For the Tungus, a nomadic tribe of hunters, the mammoth was known as Agdyan-kami (Giant Animal), a very mysterious creature because they could not explain its origin. The mere sight was believed to bring bad luck. The superstitious fear was so strong that members of the tribe became ill if they saw one. Chief Schumachoff was told off by his fellow tribesmen for meddling with the unlucky find and, after selling the tusks, he fell ill with a serious nervous disease!

The Botanist Mikhail Adams had stayed on in Siberia to continue his studies on the return journey from an unsuccessful diplomatic mission to China in 1805. He heard about the mammoth in 1806 (six years after it was first found) and rushed to the site.

By now, most of the soft parts had been fed to dogs or been scavenged by wild animals. Adams, however, managed to recover preserved parts of the head (including an eyeball) and two feet still frozen in ice, the entire skeleton and thirty pounds (approx. 14kg) of mammoth hair. These were taken back to St Petersburg and exhibited in Peter the Great's "Cabinet of Curios". The latter was to become the famous Russian Natural History Museum.

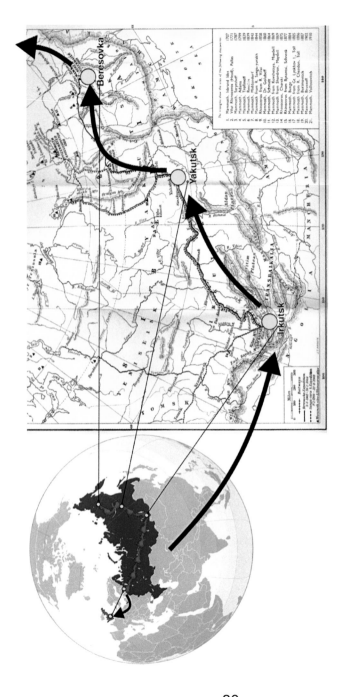

Figure 12. The 18,000 km round trip from St Petersburg to the Berezovka River where the mammoth was found
Map used – "Siberian Man and Mammoth, Pfizenmayer"

Peter the Great and later Catherine II, a German princess who became Empress of Russia, strove to bring the country out of its dark ages. They created the Russian Academy and opening the doors to scientists from other countries.

Since Adams' mammoth, and despite a longstanding offer of a cash reward, the Berezovka mammoth find in 1901was the first promising news of an intact specimen for nearly a century.

So it was that the German, Dr Otto Alfred Herz, curator of the Russian Acadamy's Zoology Department, was tasked with the expedition to retrieve the Berezovka Mammoth. Herz was a respected entomologist who had collected beetles, butterflies and moths, initially for an insect dealership in Dresden and then for the Russian Academy.

Herz took with him S. P. Sevastianov, a geologist from Yurievsk University, and the 36-year-old Eugen Wilhelm Pfizenmayer, a German taxidermist who had always dreamt of excavating prehistoric monsters. They set off on Thursday the 3rd May (based on the old Julian Calendar, 17th May in our Gregorian calendar) on what was to be an 18,000 mile round trip to the distant icy wilderness of Siberia.

The best record of the journey is E. W. Pfizenmayer's account, in his book "Siberian Man and Mammoth".

The first part of the journey was by train, via Moscow, then across the monotonous western Siberian plain to Irkutsk, just 50km west of Lake Baikal.

On the 20th May they set out in two coaches, called tarantas, each drawn by three horses. They followed the shores of the River Lena, armed and on the look-out for escaped convicts who were a real hazard as highwaymen in this region.

After three days, travelling 220 miles, they reached a steamer station at Shigolova, where the river was now navigable. They sailed to

Yakutsk, which was also a region with many political exiles deported to Siberia. They rested for 3 weeks and prepared for the remaining 1,900 mile journey on horseback.

The last major outpost they rode through was Verkhoyansk, 300 miles north, where they stopped and attended a ball! From thereon, they travelled through the Taiga and Tundra, accompanied by a Yakut guide, into Yakut territory.

They learnt a lot about local tribal customs. The Yakuts, for example, had seen the remains of many mammoths. They believed that mammoths were monsters that lived under the earth or underwater and had died when surfacing and being exposed to the light of day!

Figure 13. A Yakut couple, 1901 (photo Pfizenmayer)

On the 15th September, with snow starting to fall, they finally reached the log blockhouse that was to be their home. Too excited to rest, they rode for another half hour to where the mammoth carcass lay. They could smell it before seeing it due to the decay!

Pfizenmayer writes "We stood speechless in front of this evidence of the prehistoric world, which had been preserved almost intact in its grave of ice throughout the ages. For long we could not tear ourselves away from this primeval creature, so hung about with legend, the mere sight of which fills the simple children of the woods and tundra with superstitious dread."

Figure 14. The exposed head and front feet of the Berezovka mammoth
(photo Pfizenmayer)

They built a log blockhouse around the mammoth and gradually raised the temperature. Whenever a portion of the body had thawed sufficiently, it was carved off, investigated, and then sewn into reindeer or cowhides. It was then placed outside to quickly refreeze. Despite having taken a whole collection of different preserving methods, in the end, freezing was the simplest and the best. Both the long hair and the woolly undercoat lay all around the mammoth. Most of it had shed from the body and was also collected.

Figure 15. The Log Blockhouse at Berezovka (photo Pfizenmayer)

They discovered the mammoth had a broken front limb and broken pelvis, suggesting that it had died after falling into a crevice. It must have lain a while before finally dying as food was found, still in its mouth and stomach, with the imprint of the teeth on the cud.

On the 15th October, ten sledges towed by horses set off with the mammoth remains on the slow journey back to civilisation. With the onset of winter, they reached Kolymsk. The sun no longer rose as they continued their 3,700 mile sledge trek on the 24th November. Switching to reindeer as draught animals accelerated their progress. On 13th February, they reached Irkutsk and were able to take a train with a refrigerated mail van. They travelled by rail all the way to St Petersburg, arriving on the 18th February, 1902.

Eight days later the Tsar and Tsarina came to inspect the now very smelly, thawing carcass being prepared for exhibition. The Empress was not very impressed!

Figure 16. Eugen W. Pfizenmayer by the dogsled that took him on the first leg of the journey back to St Petersburg (photo Pfizenmayer)

A couple of months later, about a year after the expedition first set out, the skeleton was mounted as an exhibit. The most important soft parts were also preserved in alcohol for future study. Next to the skeleton, in a big glass case, the partially reconstructed skin formed the main display which is still a centrepiece to this day!

A small sample of this Berezovka mammoth hair came to the Norris Museum; a tiny part of a great adventure. How it found its way into the Norris collection is still a mystery!

Figure 17. The Norris Mammoth hair sample X.1559. The Norris Museum

Figure 18. The Berezovka mammoth as it was mounted in 1902 in the Zoological Institute Museum in St Petersburg. It is one of the museum's most valuable exhibits and can still be visited today (photo Pfizenmayer)

Hot Press! Link to "Mammoth Schmidt"

Sandra Freshney, Archivist with the Sedgwick Museum of Earth Sciences in Cambridge, came across mention of our Mammoth exhibition and the mammoth hair. Their collection has letters from another Siberian explorer and mammoth hunter from the 1870s: '...mammoth Schmidt of Siberian fame. The name suits him well, for with his long shaggy hair and beard he is not unlike the great hairy elephant whose remains he has so successfully transferred from thin icy bed to the Petersburg Museum' (Quote from Professor McKenny Hughes wife, Mary, circa 1892).

The Ice Ages

The earth has experienced several ice ages over the past 500 million years or so. We are currently in an ice age that began 2.58 million years ago. Within an ice age, there are very cold periods called glaciations where ice spreads out to cover more of the globe. During warmer interglacials, the ice recedes again temporarily (in terms of geological time). Glaciations come in waves of about 100,000 years. Currently, we are in the most recent interglacial, known as the Holocene.

In my mind, mammoths are linked to "The Ice Age", a period lovingly captured in the film series of the same name. So, I was surprised to find that, in fact, we are still in an ice age and that it is one of many.

The current series of ice ages actually began over 50 million years ago. The current and most recent ice age, the Pliocene-Quaternary glaciation, has so far lasted for about 2.58 million years. The evidence comes from fossils, geology and measurements of gases trapped in rocks and ice.

Ice ages are made up of the familiar cold periods with glaciations (the spread of ice-sheets) separated by warmer times called interglacials. Sheets of ice grew out from the North and South poles to cover substantial parts of the world and then receded again in rhythms of about 40,000 and 100,000 years.

We are currently living in an interglacial named the Holocene that began about 11,000 years ago. There are scientific debates that predict the return of a colder ice-age at anywhere between 1,500 years to 50,000 years from now. Ironically, manmade global warming may delay the onset of the next cold period.

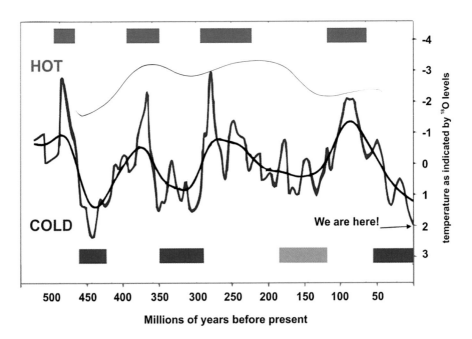

Figure 19. The Earth's hot and cold periods over the last 542 million years. The blue line is the short-term average temperature; the black line is the long-term average temperature. Graph simplified from "Phanerozoic Climate Change". Licensed under CC BY-SA 3.0 via Wikimedia Commons.

The causes of ice ages and their warmer interglacials are still an active area of scientific debate. It is an interesting area to research in itself.

Ice cover and the UK

Ukrah

Over the past half a million years, there have been five periods of glaciation. During three of these, the ice migrated south to cover substantial parts of the UK. In the last period, named the Devensian, only southern England remained free of ice up to the Wash and the Midlands.

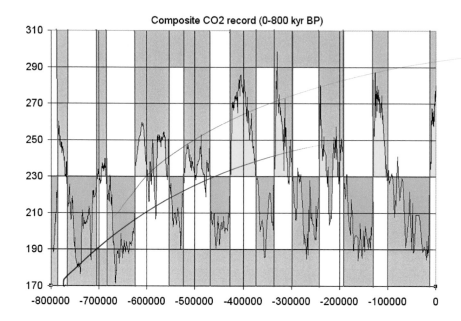

Figure 20. Cold periods (glaciations, coloured blue) and hot periods
(interglacials, coloured yellow) in the last 800,000 years of our ice age.
"CO2 glacial cycles 800k" by Tomruen. Licensed under CC BY-SA 3.0 via
Wikimedia Commons.

As the ice moved into the UK, our climate became positively freezing,
with conditions very much like Siberia in the present day. As the ice
receded from the UK, the climate became warmer. Indeed, in the
previous interglacial, the Ipswichian, temperatures in the UK were
much more like those found in the Mediterranean.

For a considerable period of time, the British Isles were connected to
Europe by a land-bridge. During glaciations, sea-levels could drop
by as much as100m!

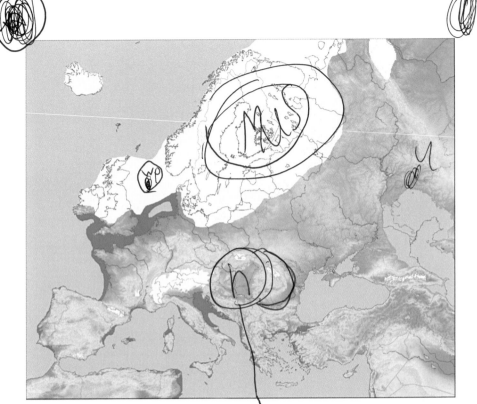

Figure 21. Ice cover over Europe and the UK during the last glaciation, the Devensian. "Weichsel-Würm-Glaciation" (another name for Devensian) by Ulamm. Licensed under CC BY-SA 3.0 via Wikimedia Commons.

So, another important factor for the past 400,000 years is that the sea levels rose sufficiently during the warm interglacials to cut Britain off from Europe. This would trap populations of animals that had migrated into Britain on the temporary island, till the climate cooled and the land bridge was re-established. It would also prevent animals and humans migrating to Britain for part of the interglacials.

This is important when we consider the mammoths and ancient elephants that came to Britain, roamed and lived in our regions.

Mammoths and Man in St Ives (and the rest of Huntingdonshire, UK)

During the waves of glaciations, when ice swept down to cover significant parts of Britain, and the warmer interglacials when the ice receded, different species of mammoths and elephants came to our island. The land bridges that existed with Europe also allowed early humans and later modern man to colonise Britain. Situated in a river valley, St Ives (and the rest of Huntingdonshire) saw mammoths, elephants and man.

"Have a look upstairs in the stores," The curators suggested, "and see what you think would look good in the exhibition." Most of any museum's collection is in storage.

I ascended the narrow steps into the attic, dodged supporting beams and found the section marked "Palaeontology". At just over six foot, I was stooping under the eaves and so sat down with relief on the low museum stool I'd bought along. To my left were fragments of mammoth tusk, Aurochs horn and a miscellany of other antlers and bones. On my right, cardboard museum boxes with smaller objects. I opened the first one and carefully took out a mammoth molar, the enamel ridges glinting in the roof light. This was the spark that made me want to learn more about mammoths and ice ages in our region and in the UK.

The evolution of mammoths and elephants

The common ancestor of elephants and mammoths was probably the Paleomastodon. They lived in Africa about 35 million years ago. Their descendants evolved and migrated across the globe.

Three million years ago, before the present Pliocene-Quaternary glaciation, there were several different elephant like or probiscidean (animals with trunks!) species living in Europe, from Anancus, to the earliest mammoths and elephants.

The Southern mammoth

Figure 22. *Mammuthus meridionalis*. The Southern mammoth on St Ives High Street. Painting & collage C. Thomas

The first mammoth to migrate into the UK and our region, 2.5 million years ago, was the Southern mammoth *Mammuthus meridionalis*. Molars of the Southern mammoth have been found in Norfolk and in the Channel, which was a land-bridge to Europe at the time.

Fossils found with the Southern mammoth show that it lived when the climate was slightly warmer than today. It browsed on leafy trees, including oak, ash and beech. At 4m height, it was one of the largest mammoth species, weighing in at 6 to 8 tons. *M. meridionalis* ranged across the whole of Europe and Asia – and into Britain via a land-bridge.

With time, although the summers might have been warmer, the winters became colder and less suitable for the Southern mammoths. Eventually, the ice swept over the landscape in the next glaciation. Southern mammoths existed until about 1.5 million years ago, so may have left these shores and returned as the ice pushed south and then retreated in 100,000 year cycles.

The Steppe mammoth

Figure 23. Mammuthus trongontherii, the Steppe mammoth. Visualised walking across the fields near St Ives. Painting and photo collage C. Thomas

As the ice receded again 600,000 years ago, a new mammoth species migrated into Britain. Its teeth show an adaptation more

33

suitable for grazing harder grasses and shrubs, prevalent in the steppes or prairies that spread across northern Eurasia. Whilst the leafy-tree-browsing Southern mammoth had 10 to 14 enamel plates on its molars to grind food, the Steppe mammoth had up to 18 plates.

We are fortunate in the UK and particularly in East Anglia, because one of the most complete skeletons of a Steppe mammoth was found in eroding cliffs near West Runton in Norfolk. It was promptly named the "West Runton Elephant" by the press. The accompanying fossil pollen and bones of frogs, newts, lizards, snakes and small mammals and birds suggested that it lived near slow moving fresh water, near the sea, with plenty of vegetation and moist woodland.

Steppe mammoths were the largest of the species, reaching heights of 4.5m. The largest bulls could have weighed 10 tons and some had curved tusks up to 5m long. Living in herds, they would have consumed enormous amounts of vegetation. A modern elephant, for example, requires up to 150kg of food a day.

The Steppe mammoth may also have been covered with fur as an adaptation to a cooler climate.

The Steppe mammoth left Britain during several glaciations and returned in the interglacials. It finally disappeared around 200,000 to 370,000 years before present. The Steppe mammoth was the forerunner of the "Ilford mammoth" and the woolly mammoth.

The Ilford mammoth and straight-tusked elephant
After a colder spell around 270 – 240,000 years ago, the UK entered the interglacial also known as MIS 7 (Marine Isotope Stage 7). The climate in the UK became positively Mediterranean. The newest elephant and mammoth migrants into our region from Europe were a small mammoth and the straight-tusked elephant.

The straight-tusked elephant was a foliage browser, preferring trees & shrubs. The small mammoth, named the Ilford mammoth after the site where it was first found, was very abundant and appeared to be a grazer, preferring more grassland.

Figure 24. Part of a jaw with a molar - probably straight-tusked elephant.

The Norris Museum, X.0260

Initially, the Ilford mammoth was thought to be a woolly mammoth, based on the enamel pattern of its molars. However, work by Katherine Scott and her team in Oxford concluded that the mammoth was more likely to be a dwarfed variety of the Steppe mammoth. This matched similar finds in Europe and the more temperate climate.

The Woolly mammoth

The woolly mammoth *M. primigenius* first appeared during the glaciation of 180,000 to 130,000 years ago. Evolved from the Steppe mammoth, this mammoth was perfectly adapted to the much colder glacial climate. It had developed a thick woolly undercoat and longer outer hair and it was padded with a layer of fat under the skin that was several inches thick and acted as insulation.

Figure 25. The smaller, chubbier and hairier Woolly mammoth, Mammuthus primigenius. Painting and photo-collage C Thomas

Woolly mammoths grazed on the hardy grasses and vegetation that survived in a fairly treeless tundra or steppe, in the region to the south of the ice sheet that covered the lowland of central England and the Welsh mountains.

The tough food had resulted in the development of molars that had many fine ridges of strong enamel to grind the abrasive grasses. The tusks were often curved and large.

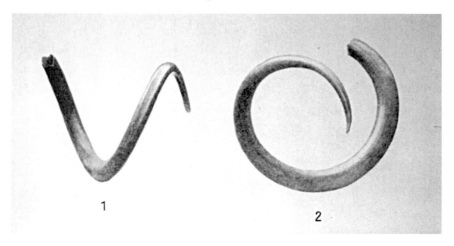

Figure 26. Woolly mammoth tusks from Siberia. From Pfizenmayer, "Siberian Man and Mammoth"

Woolly mammoths migrated out of/disappeared from our countryside during the next interglacial, the Ipswichian, from about 130,000 to 115,000 years ago. They were replaced by the straight-tusked elephant.

It is interesting that there are no records of humans in Britain during this interglacial. The reason could have been the loss of the land-bridge to Europe.

From about 400,000 years ago, Britain was regularly separated from the rest of Europe by rising sea levels that flooded the Channel.

With the onset of each ice age, the sea level would fall low enough to re-establish a land route. ~~[handwritten annotation]~~

With the end of the Ipswichian interglacial and the onset of the Devensian, the last glaciation before the present day, sea levels sank and Britain was again connected to Europe.

The woolly mammoth returned to Britain and remained until the ice receded. Again, Tundra landscapes of grasses, sedges and small shrubs provided the food for the grazing woolly mammoth.

At the peak of the last glaciation, the "Last Glacial Maximum" or LGM, it was too cold even for mammoths and they retreated from Britain and North-West Europe for the period of 21,000 – 18,000 years before present.

As the cold receded, they returned and remained on our island until at least 14,000 years ago. The warming climate and encroaching forests pushed them out of Britain, possibly across the land-bridge that was Doggerland, which still connected us to Europe until about 8,500 years ago.

During their peak, woolly mammoths were a very successful species. The Russian researcher Sergei Zimov estimates that during the last Ice Age, parts of Siberia may have had an average population density of sixty animals per hundred square kilometres - equivalent to African elephants today.

Humans and mammoths in Britain

Like the mammoths, early man also came to Britain.

Coastal erosion in Happisburgh (pronounced 'Haze-burra') in Norfolk revealed that early man was living and hunting in Britain around 800,000 years ago, when early mammoths still roamed the countryside.

Britain was a European peninsula, connected at its southern and eastern borders. Archaeologists found a string of early sites close to a now lost watercourse named the Bytham River. This would have been the earliest entry point into Britain.

Finds in Sussex at Boxgrove showed that a new human, *Homo heidelbergensis*, lived on our shores. They made distinctive oval hand axes and there was evidence from butchered bones that they hunted large mammals. There is a theory that they drove large animals (including Steppe mammoths) over cliffs or into bogs to kill them more easily.

20mm

Figure 27. Palaeolithic hand axe, Bluntisham. The Norris Museum

From then on, humans intermittently abandoned and recolonised Britain as the glaciations came and went. There is, for example, no record of humans in Britain from about 180,000 to 60,000 years ago. Several factors might account for this absence: a) it was either too cold or Britain was cut off from the continent during the warm periods or b) human populations might simply have been too low to support migration into Britain at the time.

DNA studies suggest that Britain was again periodically visited from about 41,000 years ago by modern man, who finally became established and prospered when the last glaciation receded.

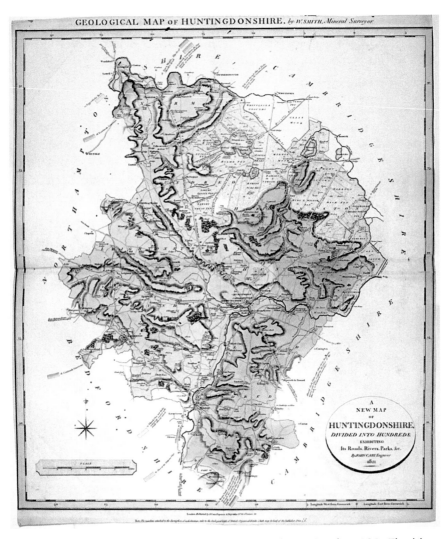

Figure 28. Geological map of Huntingdonshire by W Smith, 1821. The blue to the East comprises Alluvial clay, flints and stones. The grey to the West is mainly clay. The Fens are beige in the North-East. In the North-East, first limestone (brown) and then oolitic rock (yellow) appear.

40

Norris Museum Rocks, Records and Mammoths

The Norris Museum has marine fossils from the mudstone of the Late Jurassic period that forms the bedrock in Huntingdonshire. Glaciation scoured away most of the later deposits and then deposited clay and debris. The Great Ouse gradually developed, creating a braided river system. Here we find fossils from the warm Ipswichian interglacial and the following last Devensian glaciation. Stone tools from 300,000 years ago and from about 40,000 years ago onwards show that both old and modern humans lived in Huntingdonshire.

Over hundreds of millions of years, layers of rock formed, were buckled and eroded in our region. Presumably the scouring by the ice sheets removed most soft material in the Ouse Valley around St Ives and towards the sea. The bedrock is now Late Jurassic mudstone, dating from about 146 – 161 million years ago, when this region was under the sea. It is rich in the fossils of extinct marine vertebrates and dinosaurs, as seen in the collections of the Norris, Sedgewick and Peterborough museums.

Relating to this period, close to the Norris Museum entrance, the display includes ichthyosaur bones, large and small ammonites and models of a plesiosaur and ichthyosaur that immediately catch the eye of the visitor.

A major glaciation around 400,000 to 450,000 years before present covered the area. When the glaciers retreated, they left behind a 20 – 30m thick layer of mud, pebbles and sand called boulder clay.

Water running off the hills to the south and west resulted in the fore-runners of the River Ouse. It eroded its way through the boulder clay in the valleys. From St Neots to Huntingdon, the river valley can be up to a couple of kilometres wide. Downstream from Huntingdon, the valley broadens out into the wide Great Ouse flood plain, less than 10m above current sea level.

Over the next 250,000 years (interspersed with at least two glaciations), the ancient Great Ouse washed out the pebbles and sand as it eroded the boulder-clay, and deposited them in the river valley and in the flood plain. Around St Ives and seawards, the river took on the form of a braided river system. The river meandered over the flood plain, moving pebbles and sand from one region and creating sand and pebble banks in others.

In 1821, W. Smith, mineral surveyor, prepared a "Geological Map of Huntingdonshire" showing the clay and other features. A hand coloured print is in the Norris map collection (see figure 28).

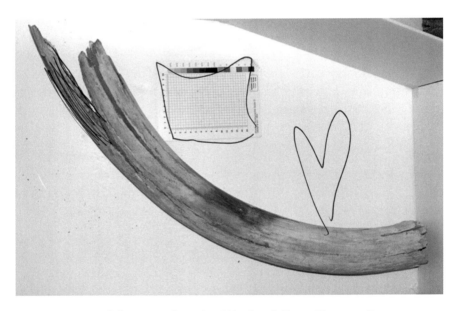

Figure 29. Tusk fragment found at Woolpack Farm Pits near Fenstanton.

The Norris Museum

The Norris Museum has a range of mammoth and straight-tusked elephant fossils. Of the 44 items, 33 came from Huntingdonshire and of these, 32 were from locations along the Great Ouse from Paxton to Earith. Most of these samples were found in gravel pits in the latter part of the 20th century. Nearly half of them (15) came from Fenstanton.

The vast majority of the samples are molars (teeth), followed by tusk fragments. Twenty-three samples were described as mammoth, six were from straight-tusked elephants.

Researchers from Cambridge University studying boreholes and the Woolpack Farm gravel pit in Fenstanton, identified two main historical periods: The last glaciation and the Ipswichian interglacial before it.

The Ipswichian interglacial included a particularly warm period around 120,000 years ago. On average, summer temperatures were 4°C warmer than today, though winters were a bit colder. This was a period when giant deer, the aggressive aurochs (wild giant cattle) as well as red and fallow deer, were preyed upon by wolves and brown bear. The Norris Collection contains several aurochs horn cores and individual bones from rhinoceros.

The main proboscidean (animal with a trunk) was the straight-tusked elephant. Despite extensive research on the Woolpack Beds, Barrington near Cambridge and a whole swathe of sites from the River Severn to the Suffolk and Essex coasts, there is no evidence of mammoths living in the UK during this warm period.

In 2000, Goa and others from Cambridge published a detailed study of microfossils, ranging from the minute bones of frogs and rodents, to beetles and other insects through to pollen and wood. These give an insight into the climate and environment in the past. Deciduous woods covered the hills, reaching down to a clearer area around the banks of the River Ouse. There is evidence of mud-churning, quite consistent with modern day drinking and wallowing sites for large herbivores in Africa. The presence of carrion eating beetles suggests that large animal carcasses were also occasionally to be found along the river's edge.

Fossils show that, when the climate cooled again during the final glaciation, Woolly mammoths roamed the valleys around St Ives. Bison and horse were also present. There were no trees and vegetation was sparse. It is thought that the gravel deposits, left behind by floods caused by summer melting of the ice, could date from 40,000 to 10,000 years ago.

The last glaciation is also the period of the human recolonization of Britain. This is reflected in the Norris Museum's collection of Palaeolithic finds. Out of 211 objects, 166 are from Huntingdonshire and most are linked to places along the Great Ouse, from Little Paxton to Earith. There are hand axes, flint scrapers and flakes.

As the last glaciers melted away and we entered the Holocene. Woolly mammoths either migrated away or disappeared. The increase in Mesolithic and Neolithic flint finds suggests that, in contrast, humans prospered and established themselves in our region.

Figure 30. Evidence of man's occupation. From left to right: Late Palaeolithic flint axe-head; Neolithic narrow flint hand axe; Neolithic ground stone axe-head. The Norris Museum

The Marvellous Mammoths Exhibition

Setting up an exhibition is both an exciting challenge and a daunting prospect! Your audience ranges from parents with very young children, the casual visitor, through to the knowledge seekers.

The Norris Museum mammoth exhibits and the research for the body of this booklet provided the ground work. As I learnt from curator Sarah Russell, you have to strike a delicate balance between text and image/display. People might not read all the information accompanying the exhibits – but they do expect it to be there.

In the end we had seven boards, each with a story or theme. The exhibits comprised the mammoth hair, the mysterious envelope, bones, tusks and teeth, alongside some Palaeolithic tools.

Assistant curator, Gilly Vose, was my mentor for, and helped with, the setting up of the exhibition. Gilly also prepared a reading and activity corner for children, linked to Neal Layton, author of the Oscar and Arabella children's books.

Helen Giles, the previous curator, had arranged the loan of a beautiful hairy mammoth model from St Albans Museum. Promptly renamed Norris, we painted an icy landscape panorama on the wall behind him. Museum assistant Mick found some eerie recordings of trumpeting mammoths to add to the atmosphere.

My wife Jane designed a giant mammoth board game and Eric, the museum's handyman, made a set of chunky wooden mammoths to use as game pieces.

We had many visitors between January and April 2015. When storyteller Marion Leeper came with her multipocketed story blanket, children were enthralled and engaged in the tale of "Hot, Hot, Hot!" Adults enjoyed the displays too and actively brought us new leads and information as a result.

Figure 31. The Exhibition boards and 'Run Mammoth Run!' game. You can see them in more detail at https://goo.gl/BjZPT9.

Figure 32. Setting up the exhibition; the final display with Norris the mammoth, visitors and friends; sticking the woolly hair on Oscar.

And Finally...

I hope that you have shared my enjoyment of the journey that began with an unassuming clump of hair – a small part of a local museum's collection. I never expected it to be linked to one of the most famous intact mammoths found. Who would have thought that we are still in one interglacial of an ongoing ice age that has shaped our countryside? Without the research that led to this book, I would not have learnt of the different species of mammoth and man that left their traces in our land.

Holding the locally excavated jawbone of a mammoth, or marvelling at the smoothness and feel of a Palaeolithic axe made by a different species of human; these are direct links to a very real past in Huntingdonshire.

It will be the same where you are. There will be artefacts with their stories just waiting for someone to dig a little deeper.

Local museums are rapidly changing – on the one hand squeezed by the paucity of money and on the other, having to reach out to the surrounding community. There will be a museum near you with items in their vast collection, which have not been seriously looked at since they were collected a century ago. They need new researchers and volunteers who can help bring your local history to life. You could be one of them.

My little book is just one story. What are you going to discover and tell us about in yours?

Let me know how you get on.

Chris

Dr Chris Thomas, chris@miltoncontact.com

Thanks

To Family, Friends and Colleagues

Who put up with my interest in the Museum and Mammoths!

The Norris Museum and Volunteers

The Norris Museum is a little gem in St Ives that is constantly juggling limited resources with the task of being the Museum for Huntingdonshire. Quite frankly, without its dedicated staff, the volunteers who give their time freely and the Friends of the Museum who support it, it could not survive.

Personal thanks to: Helen Giles for inviting me in whilst she was Curator, making me welcome and then luring me into becoming a volunteer!; Sarah Russell who, as current Curator, continued the tradition; Gilly Vose, Assistant Curator, for help with the general planning and setting up of the exhibition; Museum Assistants Richard, Mick, Lesley & Shelley; Fellow volunteers Rodney & Chris.

The Goodliff Award of the Huntingdonshire Local History Society

The Goodliff Awards were established in 1996 thanks to a generous bequest by the late Phyllis Goodliff. Each year the society makes several awards to individuals, societies, schools, museums and other institutions seeking support for projects involving the history of Huntingdonshire. This book received an award in 2015 without which it could not have been printed.

References & Resources

Investigator guides to download

Studying hair:
https://dl.dropboxusercontent.com/u/1646983/Mam
moth%20booklet/investigator-guide-to-studying-
hair.pdf

Writing and writing materials:
https://dl.dropboxusercontent.com/u/1646983/Mam
moth%20booklet/investigator-guide-writing.pdf

Gaps in the record:
https://dl.dropboxusercontent.com/u/1646983/Mam
moth%20booklet/Investigator-guides-gaps-in-the-
record.pdf

References & Links

The Norris Museum http://www.norrismuseum.org.uk/
The Goodliff Award http://www.huntslhs.org.uk/?page_id=36

Mammoth Hair

Valente A. 1983. Hair structure of the Woolly mammoth, Mammuthus primigenius and the modern elephants, Elephas maximus and Loxodonta African. Journal of Zoology. 199 (2): 271–274.
http://onlinelibrary.wiley.com/doi/10.1111/j.1469-7998.1983.tb02095.x/abstract

Hair. Wikipedia. https://en.wikipedia.org/?title=Hair

Tridico SR et al. 2014. Interpreting biological degradative processes acting on mammalian hair in the living and the dead: which ones are taphonomic? Proceedings of the Royal Society B.
http://rspb.royalsocietypublishing.org/content/281/1796/20141755

Ryder ML. 1974. Hair of the mammoth. Nature 249 (5473): 190-192.

Silvana RT et al. 2014. Megafaunal split ends: microscopical characterisation of hair structure and function in extinct woolly mammoth and woolly rhino. Quaternary Science Reviews. 83: 64-75

Envelope & Writing

Envelope. 2015. Wikipedia. https://en.wikipedia.org/wiki/Envelope

Penmanship. 2015. Wikipedia. https://en.wikipedia.org/wiki/Penmanship

Fountain pen. 2015. Wikipedia.
https://en.wikipedia.org/wiki/Fountain_pen

Saint Petersburg. 2015. Wikipedia.
https://en.wikipedia.org/wiki/Saint_Petersburg

Mammoth Rescue 1901/1902

Pfizenmayer EW. 1939. Siberian Man and Mammoth. London & Glasgow. Blackie & Son Limited. Translated from the German by M. D. Simpson.

Ice Age

Ice Age. 2015. Wikipedia. https://en.wikipedia.org/wiki/Ice_age

Phanerozoic Climate Change". 2010. Wikimedia Commons. https://commons.wikimedia.org/wiki/File:Phanerozoic_Climate_Change.png

Co2 glacial cycles 800k. 2011. Wikimedia Commons. https://commons.wikimedia.org/wiki/File:Co2_glacial_cycles_800k.png

Weichselian glaciation. 2015. Wikipedia. https://en.wikipedia.org/wiki/Weichselian_glaciation

Mammoths & Man in Huntingdonshire

Evolution. 2004. http://www.sanparks.org/parks/kruger/elephants/about/evolution.php

Lister A, Bahn P. 2007. Mammoths: Giants of the Ice Age. Struik

Mammoth. 2015. Wikipedia. https://en.wikipedia.org/wiki/Mammoth

Cromer Museum. (date unknown). The West Runton Elephant. Cromer Museum Brief History guide no: 17. http://www.museums.norfolk.gov.uk/view/ncc123705

Scott K. 2007. The ecology of the late middle Pleistocene mammoths in Britain. Quaternary International. 169-170: 125-136 http://www.sciencedirect.com/science/article/pii/S1040618206002382

Zimov S. 2014. Zimov's Manifesto. The Woolly Mammoth Revival. http://longnow.org/revive/projects/woolly-mammoth/sergey-zimovs-manifesto/

Prehistoric Britain. 2015. Wikipedia. https://en.wikipedia.org/wiki/Prehistoric_Britain

Gao C et al. 2000. Last interglacial and Devensian deposits of the River Great Ouse at Woolpack Farm, Fenstanton, Cambridgeshire, UK. Quaternary Science Reviews.19: 787-810 http://www.sciencedirect.com/science/article/pii/S0277379199000281